COMPUTERS
from Then to Now

BY ANGIE SMIBERT

SEQUENCE

AMICUS | AMICUS INK

Sequence is published by Amicus and Amicus Ink
P.O. Box 227, Mankato, MN 56002
www.amicuspublishing.us

Copyright © 2022 Amicus. International copyright reserved in all countries. No part of this book may be reproduced in any form without written permission from the publisher.

Library of Congress Cataloging-in-Publication Data
Names: Smibert, Angie, author.
Title: Computers from then to now / by Angie Smibert.
Description: Mankato, MN : Amicus, 2022. | Series: Sequence | Includes bibliographical references and index. | Audience: Ages 7-10 | Audience: Grades 2-3 | Summary: "Full-color photographs and a running timeline illustrate important developments in the history of computers. Topics feature the transition from vacuum tubes to transistors to microchips, computer languages and software, history of the internet, and artificial intelligence. A table of contents, glossary, further resources, and an index are included"-- Provided by publisher.
Identifiers: LCCN 2019036296 (print) | LCCN 2019036297 (ebook) | ISBN 9781681519609 (library binding) | ISBN 9781681526072 (paperback) | ISBN 9781645490456 (pdf)
Subjects: LCSH: Computers–History–Juvenile literature. | Computer science–History–Juvenile literature.
Classification: LCC QA76.23 .S565 2022 (print) | LCC QA76.23 (ebook) | DDC 004.09–dc23
LC record available at https://lccn.loc.gov/2019036296
LC ebook record available at https://lccn.loc.gov/2019036297

Editor: Gillia Olson
Designer: Aubrey Harper
Photo Researcher: Bridget Prehn

Photo Credits: Alamy/Pictorial Press Ltd 8, sandy young 22–23, Science History Images 4 (inset), 10–11, 15; AP/Auction Team Breker 19; Getty/Andrew Burton 16–17, Christian Science Monitor 6–7, James Leynse 26–27; iStock/Sjo cover, iStock/KOHb circuit texture (throughout); Science Source/NASA 20; Shutterstock/NanamiOu cover (background); paparazzza 28–29, Science and Society 24; Wikimedia/Bequest of Madame Prosper Faugère 4, John Ambrose Fleming 9, Windell Oskay 12–13

TABLE OF CONTENTS

The First Computers   5

Big to Small   9

Getting Personal   18

Today and the Future   29

■ ■ ■ ■ ■

Glossary   30

Read More   31

Websites   31

Index   32

Blaise Pascal
invents the Pascaline.

1642

The First Computers

People have always needed to keep track of things. We counted sheep in the fields. We kept track of wheat sold at the market. Over time, we made machines to do the math. In 1642, Blaise Pascal invented an adding machine. He called it the Pascaline. Early adding machines led to the first computers.

Blaise Pascal invented the Pascaline (inset).

In 1832, British inventor Charles Babbage began to build the first **computer**. He called it the Difference Engine No. 1. In his day, people used printed tables of numbers to do complex math. These tables often had errors. Babbage designed his computer to do this as a machine. However, he never finished his computer.

Babbage's Difference Engine would have looked like this! Doron Swade runs the copy he built in 2008.

Blaise Pascal invents the Pascaline.

1642

1832

Charles Babbage starts to build Difference Engine No. 1.

OADING...LOADING...LOADING...

7

John Fleming invented the vacuum tube to improve radio signals, but it would change electronics forever.

Blaise Pascal invents the Pascaline.

John Fleming invents the vacuum tube.

1642 1832 1904

G...LOADING.

Charles Babbage starts to build Difference Engine No. 1.

Big to Small

In 1904, John Fleming invented the **vacuum tube**. This glass tube had no air or moving parts inside. Electric **current** going through it could be switched on or off very quickly. This on-or-off system was **binary code**. On is one. Off is zero. In 1938, German engineer Konrad Zuse built the first computer with vacuum tubes. He built the Z1 in his parents' living room.

vacuum tube

Vacuum tubes made computing faster. In 1943, British engineers began using Colossus. It broke military codes. Each Colossus had as many as 2,500 vacuum tubes. ENIAC was a famous early computer. It was 1,000 times faster than earlier ones. ENIAC was huge. The computer used 17,468 vacuum tubes.

> ENIAC programmers had to move wires around so the computer could solve different problems.

- **Blaise Pascal invents the Pascaline.** — 1642
- **Charles Babbage starts to build Difference Engine No. 1.** — 1832
- **John Fleming invents the vacuum tube.** — 1904
- **Colossus begins operation.** — 1943

OADING...LOADING...LOADING...

Blaise Pascal invents the Pascaline.		John Fleming invents the vacuum tube.		The transistor is invented.
1642	1832	1904	1943	1947
	Charles Babbage starts to build Difference Engine No. 1.		Colossus begins operation.	

Vacuum-tube computers were huge. The tubes were big, hot, and expensive. They burned out easily. In 1947, Bell Laboratory invented something much smaller. It was called a **transistor**. The transistor could do everything a vacuum tube could but better. This little invention changed electronics and computers.

A plastic C-shaped frame holds the first transistor.

People write **programs** to tell computers what to do. The first programs were written in binary code. They used only ones and zeros. Computer scientists wanted to use words. An early word-based language was Short Code in 1949. It was slow. In 1954, IBM invented FORTRAN. It became popular. Other languages soon followed.

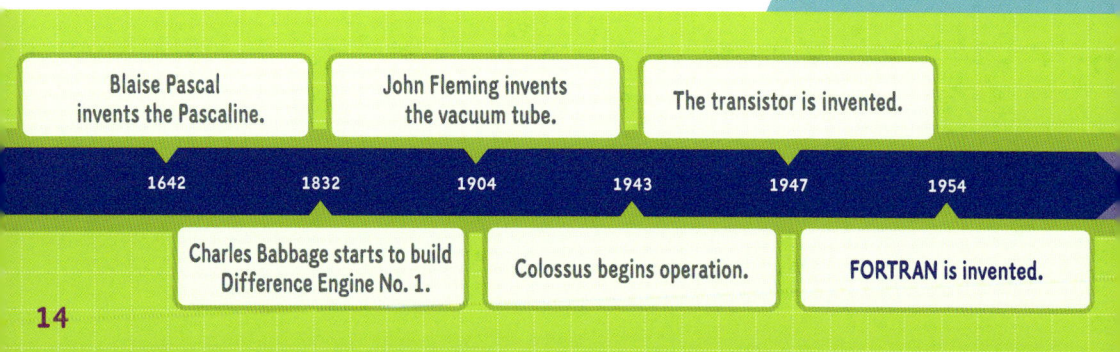

1642	1832	1904	1943	1947	1954
Blaise Pascal invents the Pascaline.	Charles Babbage starts to build Difference Engine No. 1.	John Fleming invents the vacuum tube.	Colossus begins operation.	The transistor is invented.	FORTRAN is invented.

Grace Hopper helped develop the computer language COBOL, which came out soon after FORTRAN.

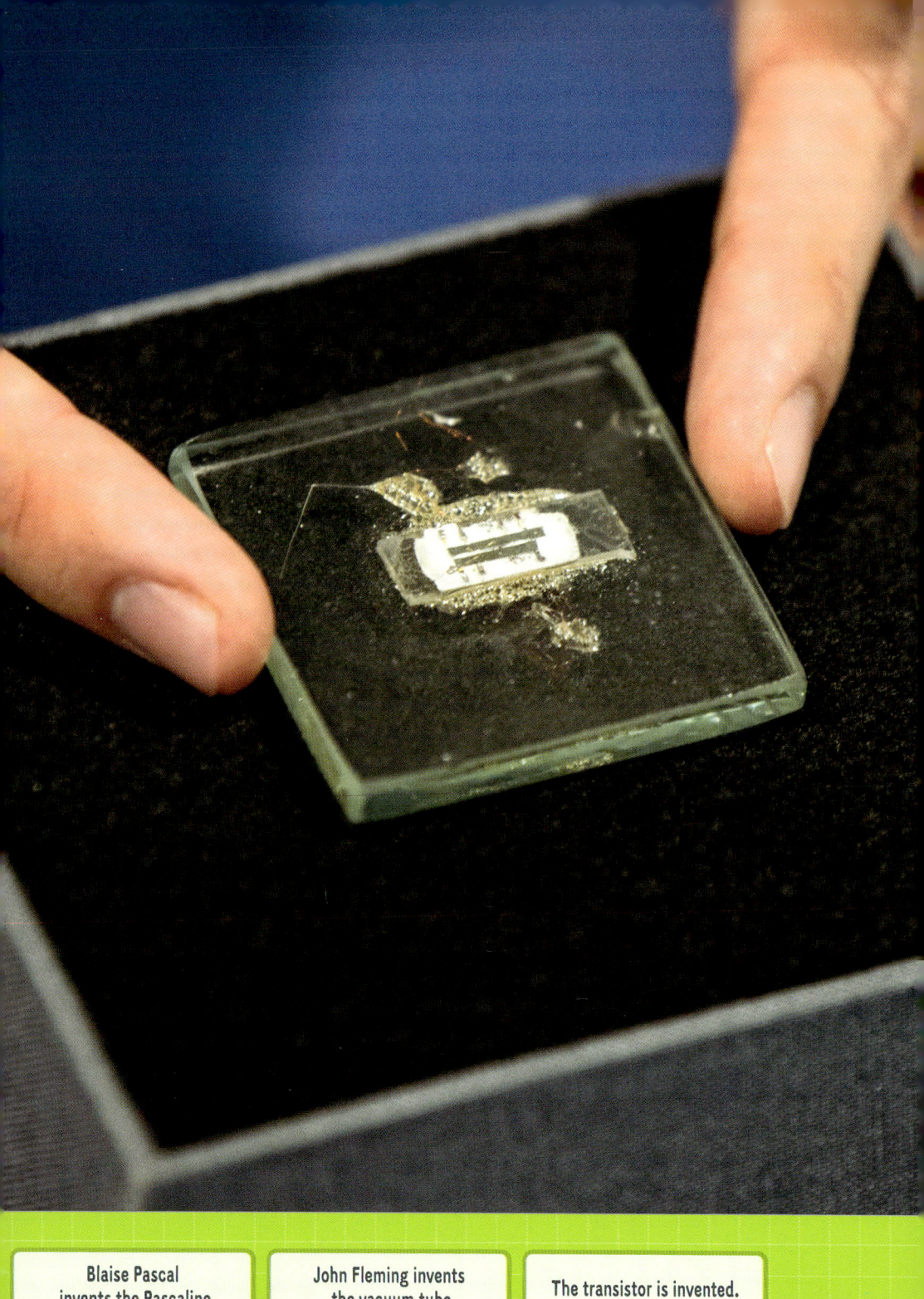

Blaise Pascal invents the Pascaline.		John Fleming invents the vacuum tube.		The transistor is invented.	
1642	1832	1904	1943	1947	1954
	Charles Babbage starts to build Difference Engine No. 1.		Colossus begins operation.		FORTRAN is invented.

16

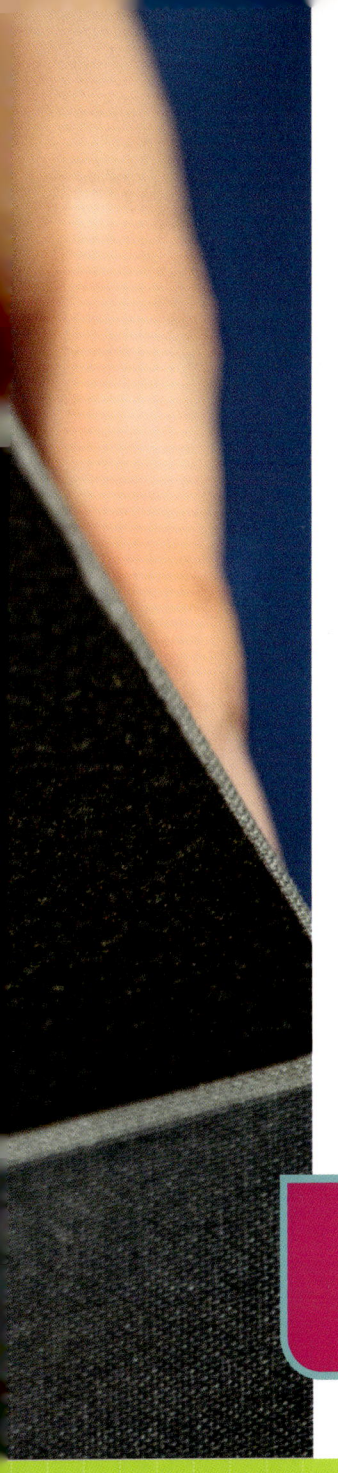

In 1958, engineers found a way to make transistors even smaller. Jack Kilby and Robert Noyce each built a **microchip**. Both had super tiny transistors. They were too small to see with just your eyes! Microchips made small computers possible. In 1969, the Apollo spacecraft used a microchip computer to land on the Moon.

Jack Kilby's 1958 microchip paved the way for smaller computers.

1958

LOADING... LOADING...

The microchip is invented.

Getting Personal

In the 1970s, computers got smaller and easier to get. In 1974, the Altair 8800 kit came out. People could build their own "personal" computer. It inspired Steve Wozniak to make the Apple I. Wozniak and Steve Jobs soon started Apple computers. In 1976, they sold 200 of the Apple I. Apple quickly came out with the Apple II. It was the first ready-to-use personal computer. They sold 1 million of them.

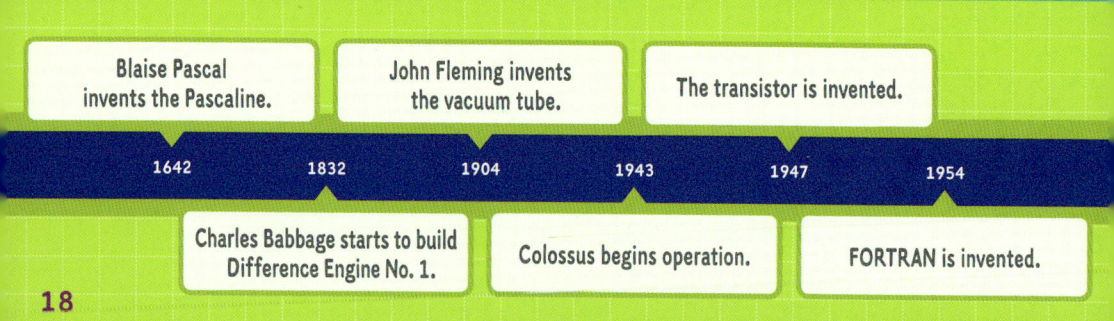

1642	1832	1904	1943	1947	1954
Blaise Pascal invents the Pascaline.	Charles Babbage starts to build Difference Engine No. 1.	John Fleming invents the vacuum tube.	Colossus begins operation.	The transistor is invented.	FORTRAN is invented.

The Apple I came with just the green circuit board. The user put together the rest.

The first personal computer, Apple I, is released.

1958   1976

The microchip is invented.

Blaise Pascal invents the Pascaline.		John Fleming invents the vacuum tube.		The transistor is invented.	
1642	1832	1904	1943	1947	1954
	Charles Babbage starts to build Difference Engine No. 1.		Colossus begins operation.		FORTRAN is invented.

20

People wanted to travel with their computers. In 1982, Grid came out with one of the first laptops. It was called the GRID Compass. It weighed only 10 pounds (4.5 kg). Its case was strong. NASA even started using it on space shuttles. GRID also made one of the first tablet computers in 1989.

Astronaut Thomas Jones uses a GRID laptop during a 1994 space shuttle mission.

1958 — The microchip is invented.
1976 — The first personal computer, Apple I, is released.
1982 — GRID releases one of the first laptops.

...LOADING...

By the 1980s, more people had computers. But they were not easy to use. People had to know how to write commands to make the computer work. In 1984, Apple came out with the Macintosh. It had both a mouse and a graphical user interface (GUI). A GUI shows programs as pictures. People click on the pictures to run the programs. It was a huge success.

Apple's GUI for the Macintosh was black and white. Color would come later.

| Blaise Pascal invents the Pascaline. | John Fleming invents the vacuum tube. | The transistor is invented. |

1642   1832   1904   1943   1947   1954

| Charles Babbage starts to build Difference Engine No. 1. | Colossus begins operation. | FORTRAN is invented. |

	The first personal computer, Apple I, is released.		Apple releases the Macintosh.	
1958	1976	1982	1984	
The microchip is invented.		GRID releases one of the first laptops.		

23

| Blaise Pascal invents the Pascaline. | | John Fleming invents the vacuum tube. | | The transistor is invented. | |

1642 — 1832 — 1904 — 1943 — 1947 — 1954

| | Charles Babbage starts to build Difference Engine No. 1. | | Colossus begins operation. | | FORTRAN is invented. |

The internet was born in 1969. Governments and universities were the only users. Plus, it was hard to use. In 1989, Tim Berners-Lee invented the World Wide Web. In 1993, the code became free for anyone to use. It connected people around the world. People had a new way to share information.

> Tim Berners-Lee poses with a World Wide Web page.

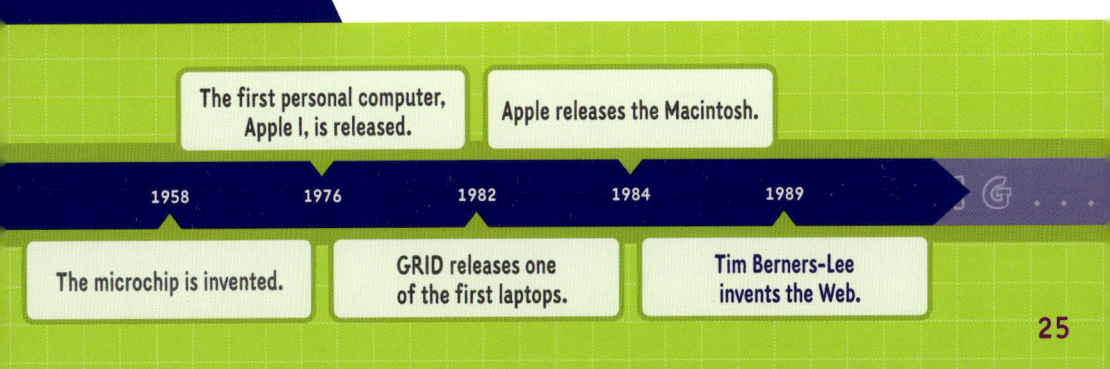

1958	1976	1982	1984	1989
The microchip is invented.	The first personal computer, Apple I, is released.	GRID releases one of the first laptops.	Apple releases the Macintosh.	Tim Berners-Lee invents the Web.

Computer chips kept getting smaller. Small chips turned cell phones into smart phones. They became small computers. The Simon phone from IBM came out in 1994. But you couldn't use the internet. In January 2007, Apple announced the iPhone. You could use the internet. It put computers in our pockets. Today, watches, cars, and smart speakers all use microchips.

Excited customers check out the iPhone on its first day of sale in June 2007.

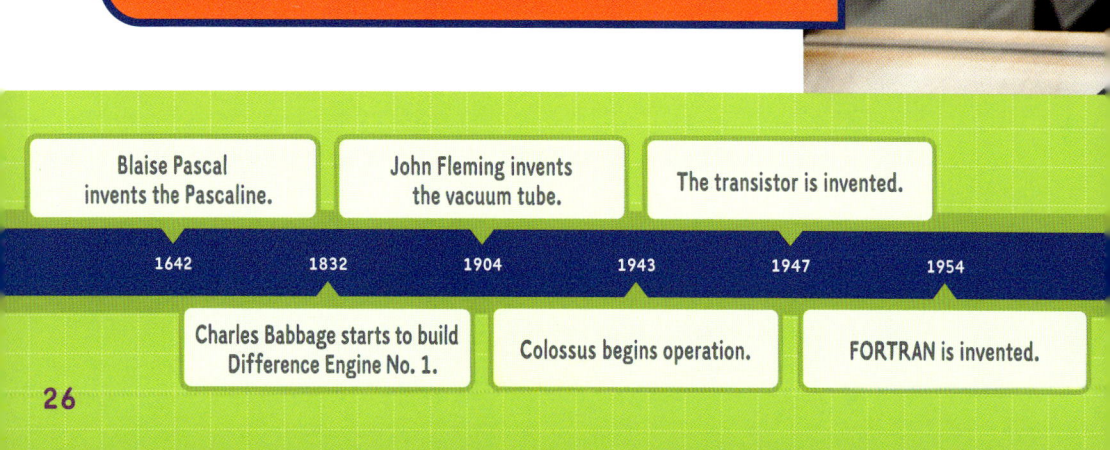

Blaise Pascal invents the Pascaline.	John Fleming invents the vacuum tube.	The transistor is invented.
1642	1832	1904	1943	1947	1954

| Charles Babbage starts to build Difference Engine No. 1. | Colossus begins operation. | FORTRAN is invented. |

	The first personal computer, Apple I, is released.		Apple releases the Macintosh.			The iPhone is released.	
1958	1976	1982	1984	1989	2007
The microchip is invented.		GRID releases one of the first laptops.		Tim Berners-Lee invents the Web.	

27

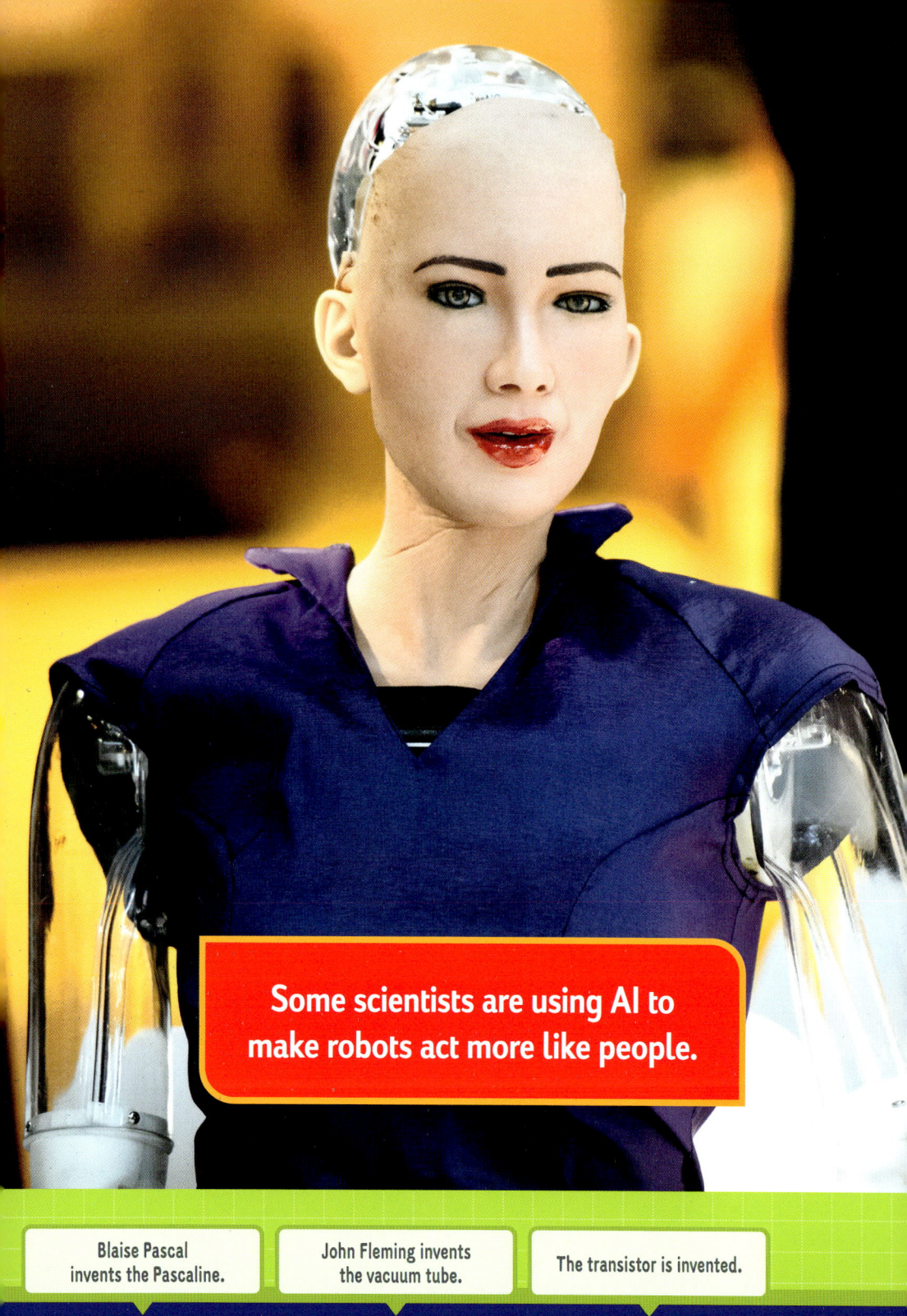

Some scientists are using AI to make robots act more like people.

| 1642 | 1832 | 1904 | 1943 | 1947 | 1954 |

- Blaise Pascal invents the Pascaline.
- Charles Babbage starts to build Difference Engine No. 1.
- John Fleming invents the vacuum tube.
- Colossus begins operation.
- The transistor is invented.
- FORTRAN is invented.

28

Today and the Future

Computers keep getting more powerful. **Artificial intelligence** (AI) takes computers in new directions. AI makes computers think, act, or learn almost like people. AI already beats the best human players in complex games. Someday, it will run self-driving cars. Computers of the future have endless possibilities.

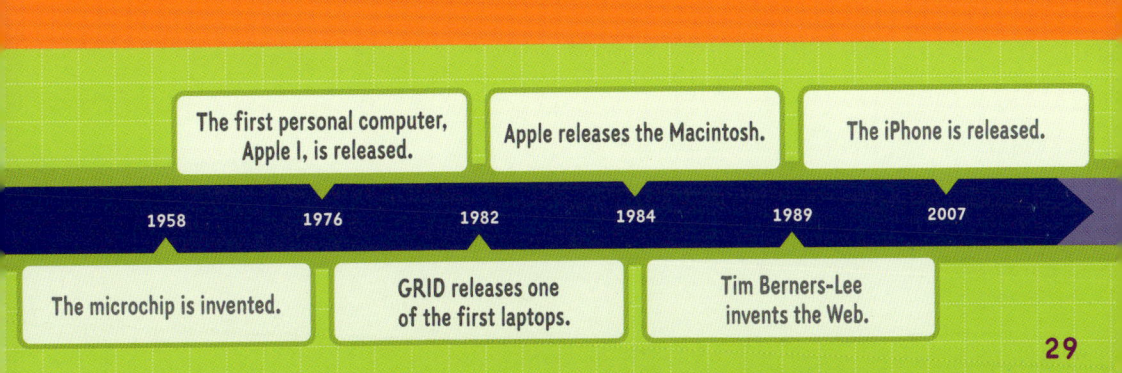

- 1958 — The microchip is invented.
- 1976 — The first personal computer, Apple I, is released.
- 1982 — GRID releases one of the first laptops.
- 1984 — Apple releases the Macintosh.
- 1989 — Tim Berners-Lee invents the Web.
- 2007 — The iPhone is released.

Glossary

artificial intelligence A computer or program that can do tasks that normally require human intelligence, such as seeing, recognizing speech, learning, and making decisions.

binary code A code based only on the numbers 0 and 1, used as the basis for computer programs.

computer An electronic machine that can store and work with large amounts of information.

current A flow of electricity.

microchip A group of tiny electronic devices that work together on a very small piece of hard material.

program A set of instructions that tell a computer what to do.

transistor A small device that is used to control the flow of electricity.

vacuum tube A glass tube that was previously used to control the flow of electricity before transistors.

Read More

Grack, Rachel. *Wireless Technology from Then to Now.* Sequence Developments in Technology. Mankato, Minn.: Amicus/Amicus Ink, 2020.

Hubbard, Ben. *How Computers Work.* Read and Learn: Our Digital Planet. North Mankato, Minn.: Heinemann Raintree, a Capstone imprint, 2017.

Smibert, Angie. *Inside Computers.* Inside Technology. Minneapolis, Minn: Core Library, an imprint of Abdo Publishing, 2019.

Websites

DK Findout: Computer Coding
www.dkfindout.com/us/computer-coding

The Kids Should See This: How Does the Internet Work?
https://thekidshouldseethis.com/post/26674356049

Ted-Ed: Inside Your Computer
https://ed.ted.com/lessons/inside-your-computer-bettina-bair

Every effort has been made to ensure that these websites are appropriate for children. However, because of the nature of the Internet, it is impossible to guarantee that these sites will remain active indefinitely or that their contents will not be altered.

Index

adding machines 5
Apple computers 18, 22, 26
artificial intelligence 29
Babbage, Charles 6
Bell Laboratory 13
Berners-Lee, Tim 25
binary code 9, 14
cell phones 26
Colossus computer 10
computer languages 14
ENIAC computer 10
first computer 6
Fleming, John 9
FORTRAN 14
graphical user interface (GUI) 22
Grid Compass 21
internet 25
iPhone 26
Jobs, Steve 18
laptops 21
Macintosh 22
microchips 17, 26
mouse 22
Pascal, Blaise 5
personal computers 18
programs 14, 22
smart speakers 26
spacecraft 17, 21
tablet computers 21
transistors 13, 17
vacuum tubes 9, 10, 13
World Wide Web 25
Wozniak, Steve 18
Zuse, Konrad 9

About the Author

Angie Smibert has written dozens of educational titles just like this one. She was a science writer and online training developer at NASA's Kennedy Space Center for many, many years. She received NASA's prestigious Silver Snoopy as well as several other awards for her work.